Keep

C000143413

A pocket guide to climate action

Brie Gyncild

First edition: 2022

ISBN 978-1-7348648-0-9

Wordy Folks Publishing

1407 15th Ave

Seattle, WA 98122

www.wordyfolks.com

Cover design uses image by Turkkub from Flaticon.com

For everyone who can imagine a better future
and is helping to bring it to life.

Contents

Introduction

We're at a critical juncture.

A couple of hundred years ago, humanity launched a grand experiment, motivated by the desire for a better quality of life. We learned how to harness the remains of things long-dead to fuel homes, factories, vehicles, and everything else in our world. In the process, a small number of people were able to become very wealthy.

Over time, we came to understand that our use of—and dependence on—fossil fuels threatened the livability of our climate. But by that time, moving away from fossil fuels was *hard*. Far too many corporations, governments, industries, and individuals prioritized convenience over sustainability, efficiency over safety, and profit over community. Together, they (we) are warming our environment, polluting our air and water, and threatening our very existence. And let's be clear—the damage isn't equally distributed. Throughout our country and the world, it's poorer communities, Black and Brown families, people with disabilities, indigenous peoples, and other disenfranchised groups who suffer the most and the earliest from changes in the climate and from business as usual.

Unfortunately, we can't go back and change decisions that were made in previous decades. But we can—and must—make better decisions with the information and technologies available to us now. Each of us has the power to make better choices about every aspect of our lives, and working together, we can change our institutions and industries.

We can't wait for governments to act, and we can't wait for industries to change. We need to pressure those entities *while* doing what we can in our own lives, families, and communities.

A UN Emissions Gap report says that 70% of total global emissions are the direct results of personal lifestyle and purchasing decisions. Seventy percent!

Of course, many of those decisions aren't truly decisions. With a society that limits options and makes it difficult to make climate-friendly choices, many people have limited transportation, food, or other choices. We can, however, work to change systems so that the climate-friendly choices are the easiest and most sensible ones as well.

Where we do have options and privilege, we can make the better choices now. We can play an important role in creating the world we want to live in.

Living a climate-conscious life is good for the planet and the species that call it home, but it's also personally rewarding. It's very satisfying to know that you're creating a more livable future when you plan vacations closer to home, take the bus instead of driving, or choose a more durable product. It can even be fun to solve the sometimes challenging puzzle of how to do something more sustainably.

Beyond that immediate gratification, many strategies for cutting greenhouse gases result in healthier bodies, stronger communities, greater equity, lower costs, greater mindfulness, changing social norms, and the inspiration for climate activism. That's a lot of benefit from walking or biking, eating vegetarian meals, skipping a flight or cruise, or line-drying laundry!

The climate crisis provides an opportunity to remake ourselves, our communities, and our economies. We can and should follow the lead of the communities who are most affected, address racial and socioeconomic inequities, and build caring institutions that put people and other living things ahead of profit.

I'm not perfect, and you probably aren't, either. But each of us can make small—and not so small—changes in our lives and our communities to make a real difference. Generally, the greatest impact we can make in our individual lives comes from finding alternatives to flying; choosing to walk, bike, roll, or use transit instead of driving when we move through the world; planning our families so that every child born is wanted and cherished; and eating a mostly plant-based diet.

The larger impact of the changes we make in our personal lives is the influence those changes have on social norms, pressure for policy changes, and exposing the challenges we face in trying to live our lives sustainably. The ultimate goal is to create systems and policies so that the easiest thing to do is also the most environmentally advantageous, whether you're thinking about the climate effects or not.

With this book, I've pulled together ways that regular people can make a difference so that you can see many options (and the reasoning behind them) in one place, so you're able to weigh for yourself where to put your climate action energy. This book is far from exhaustive, but I hope it will help you find opportunities to meaningfully effect change and create a better world.

As one of my favorite climate writers, Eric Holthaus, says in his newsletter, *The Phoenix*: "We are in a climate emergency. And you were born at just the right moment to help change everything."

How to use this book

This book is meant to be a useful resource to help you identify ways that you can take meaningful action. Reading it all at once could easily be overwhelming. Instead, choose a chapter that interests you, think about how you might make a change in your life or work with others to change policy in that area, and give it a try. (If you're not sure where to start, consider the bolded action in each section, as it's often the most impactful.) Once you've mastered one change, dip back in to find another action.

Use the note pages at the end of the book to jot down ideas as you have them. Noting what inspires you will make it easier to remember which actions you want to take, things you want to learn more about, and ideas you want to share.

Consider finding a climate action buddy or a climate action group, and challenging and supporting each other to make changes in your personal lives and in the organizations where you have influence. In many cities, there are climate change meetup groups, 350.org action groups, or other gatherings of people you can join forces with. You can also find friends virtually through communities like the Soapbox Project. With a supportive friend or three, it's easier to learn how to talk about climate change, commit to shifting behaviors in our own lives, and advocate for changes in your city, state, or local industries and governments.

You'll find resources for learning more and taking action at ClimatePocketGuide.com. It's also where you'll find sources for the statistics I cite in this book.

Speaking up and speaking out

One of the most important things you can do to curb climate change is to talk about it.

Talk with—and listen to—people in your family, your workplace, and your community about climate change and its effects, as well as the ways in which we can all make a difference. Often, the people we personally know and respect have much greater influence on us than celebrities, government officials, or statistics.

People who know that their friends and family members are concerned about climate change are more likely to support policies that address it. And elected leaders need to know that their constituents support smart, equitable climate policy.

Unfortunately, climate change has been politicized to the point that many of us are hesitant to talk about it. But chances are good that you'll discover that your neighbors and friends are also concerned about the ways the climate is changing.

Talk about it

We're seeing the effects of climate change all over the globe, and we need to act decisively. Conversation can inspire action.

- **Start the conversation.** Talk about how weird the weather is, about your fears and your hopes. Talk about specific things everyone can do (show them this book!) and about big-picture changes we need to make as a society.

- Educate yourself. Read trustworthy reporting on climate science. Watch science-based documentaries. Visit informative websites such as drawdown.org. Visit ClimatePocketGuide.com for a list of reputable sources for climate news and information.

- Ask questions and listen. Everyone's concerns are different. Start a difficult conversation by asking an open-ended question and then listen to the observations, fears, and hopes someone shares with you. Lead with emotions and human connection, and follow up with facts and information about what they can do to help.

- Practice! If you're uncomfortable talking about climate change, practice with a trusted friend or in a friendly climate-action community such as the Soapbox Project (soapboxproject.org).

- Inspire others. When you take climate action, let folks know! You don't have to shove it in people's faces, but don't be shy either. You're a role model—and you may find out they're doing incredible things you don't know about.

Elect climate-focused leaders

Elected officials at every level of government can help curb climate emissions and improve resilience to climate change. Though many people vote only in presidential elections, it's often the local races that have a greater effect on your life and the ability of your community to make effective changes.

- **If you're eligible, register to vote.** If you've missed any elections or moved, confirm that your registration is active and accurate.

- Be an informed voter. Research candidates and learn about the issues in local, regional, state, and national elections.

- Vote!! Make sure to follow the requirements to ensure your vote counts. If you're submitting a ballot through the mail or a dropbox, make sure to sign it.

- Support candidates who have good climate plans, and let all candidates know that you expect climate action to be a high priority.

Go further

- Help register voters.

- Share your priorities and endorsements with friends and family.

- Support efforts to make voting accessible and fair.

Advocate

Advocacy means telling people with power what you think they should do. Many of the actions in the Go Further sections in this book suggest supporting specific policies. Usually, that means sending an email or making a phone call to an elected official or the decision-maker in an organization or company.

- **Join local advocacy groups.** Most communities have a local chapters of national organizations, as well as groups specifically created to protect your local waterways, forests, parks, farmlands, or vulnerable communities. Get on their mailing lists to learn more about climate issues in your area and how you can get involved. (ClimatePocketGuide.com has information about local, regional, and national advocacy groups.)

- Follow the local news to learn about proposed policies and laws that might have an effect on the climate—good or bad.

- Email, call, or meet in person with local, regional, state, or national elected officials to urge them to take action. It's often more effective to advocate for or against a specific policy, but they also need to know more generally that you prioritize climate action.

Protest and rally

As we make individual choices and work to keep our communities safe, there are still far too many people invested in the status quo. And the status quo is literally killing us. Sometimes we have to get loud and disruptive to turn things around. There are many opportunities to speak out with others. Most require only that you show up—virtually, through social media, or in person. Some may result in arrest for civil disobedience. Different people have different comfort levels and different risk factors—get involved at whatever level is appropriate for you.

- **Join climate marches and rallies.** If there isn't one planned in your community the next time a national or global climate march or strike is called, organize one!

- Join Fridays for the Future, the effort that started with one teenage girl on a school strike to get attention for the need for climate action. Now there are Fridays for the Future groups protesting weekly all over the world.

- Support Water Protectors, especially indigenous activists who are fighting to keep toxic gas pipelines from defouling their sacred lands and water sources. If you can put your body on the line, join them. If that's not realistic for you, support them financially and by amplifying their message.

Use social media

It's easier than ever before for individuals to get a message out to the masses. While communicating via social media isn't enough on its own, it's a powerful complement to everything else you're doing. Amplify the voices of people who are most affected by climate change, activists who are calling for change, and inspiring people (like you!) who are making a difference in their own lives and in their communities.

- **Follow the organizations, journalists, and activists you respect.** Share their messages with your own followers.

- Push back against greenwashing from fossil fuel companies and other polluters. Check out tweets from @RealHotTake.

- Tag your elected leaders—local, state, or federal—when you share news or other people's posts that you think they should see.

Advocate for a Just Transition

It's not enough to reduce carbon emissions and curb climate change. Both the future we create and the way we get there need to be just and equitable. That means centering the communities who are most at risk and who have already suffered the most.

The Just Transition Alliance (jtalliance.org) describes "Just Transition" as a principle, a process, and a practice:

*   The principle is that a healthy economy and clean environment for all should co-exist; we don't need to sacrifice one for the other.

*   The process is a fair one, in which both workers and community residents retain (or increase) their health, environment, jobs, and economic assets.

*   The practice is that the people who have been most affected by pollution and who are most at risk should be centered and lead in the efforts to create policy solutions.

When you learn about policy proposals or industry changes, ask who's at the table. Communities most affected, including workers whose jobs will disappear during a clean-energy transition, must be engaged in policy discussions.

Think critically about how investments are made. Who can take advantage of tax incentives, for example? Are communities with greater resources better able to jump through bureaucratic hurdles to access programs? Let policymakers know you expect affected communities to advise how investments should be structured.

When in doubt, follow the lead of local and regional environmental justice groups.

Getting around

We evolved as bipeds, walking and running and climbing and jumping. For most of us, our legs work pretty well, but eventually we wanted to go farther, faster than they could take us.

We rode horses, camels, burros, donkeys, and elephants, asking them to expend their energy to bear the burdens we put on them. We invented the wheel and attached it to frames to make wagons, bicycles, tricycles, and other devices that could be powered by our bodies and the animals we tamed.

Eventually, we used stored carbon from ancient plants to power engines attached to metal boxes to move us faster with almost no effort from us—a significant achievement in propulsion but one that came at great cost to a habitable climate, our physical health, and our communities.

Now, we have more options for getting around than ever before. Walking, rolling, biking, and transit are the most efficient and fun ways to go short distances, and they let us interact with our communities as we travel. Even when it's impractical to walk, bike, or roll, or when transit isn't available, everyone can do *something* to reduce their greenhouse gas emissions as they move through their day.

Rethinking how we move our bodies through the world can not only reduce the impact on the climate, but improve our physical health, build stronger connections with our neighborhoods, save money, and make us happier!

Walk, bike, and roll!

The more we move using our own power, the more climate-friendly our transportation.

Nearly half (46%) of daily trips are three miles or shorter. For most people, that's an hour to walk or to roll in a wheelchair, or about a twenty-minute bike ride.

- **Walk or bike to work or school!** You're more likely to get there feeling energized and ready for the day—and you don't have to worry about getting stuck in traffic. Test your route on a weekend when you can take your time to find the best way.

- Get creative. Can you use a handcart to haul groceries or garden supplies? Kayak to work? Add a child seat to your bicycle? There are many ways to get yourself and whomever/whatever you're taking with you to your destination.

- Learn about resources for biking in your home city and state. Visit League of American Cyclists at bikeleague.org/bfa, People For Bikes at www.peopleforbikes.org/locations, or type "bike map" with the city name in your browser.

Go further

- Advocate for sidewalks, pedestrian-friendly signal timing, and lower speed limits to make it safer and more comfortable for people to walk.

- Learn about Americans With Disabilities Act (ADA) requirements, and demand that your local governments comply.

- Support zoning that promotes walkable communities so that people can walk to the grocery store, movie theater, school, work, doctor's office, bank, church, library, community centers, parks, and other daily destinations.

- Learn about and join efforts to make it safer and easier to walk at AmericaWalks.org.

- Support efforts to reduce violence, police harassment, and hate crimes to make walking safe and comfortable for people of color, LGBTQ folks, and people who don't have housing.

- Let city, county, and state officials know that you want safe biking infrastructure.

Use transit

In our modern world, we often have to go farther than is practical to walk, roll in a wheelchair, or bike—and sometimes steep hills, extreme weather, or physical challenges make those options unfeasible. When that's the case, choose public transit: travel by bus, train, subway, ferry, or streetcar. Though many transit fleets still rely on diesel fuel, it's far more efficient to move dozens of people in one vehicle than it is for everyone to drive separately. As fleets electrify and grids become cleaner, transit is increasingly climate-friendly.

- **Use transit to travel through your city or region.** Check local websites or kiosks for schedules and rider information.

- Use trains or buses to travel longer distances. Check Amtrak, Greyhound, BoltBus, or other companies for options. Several websites let you search multiple bus lines, and some cities have specific sites for transportation options.

Go further

- Urge your local elected officials and Congressional representatives to fund public transit investments, including high-speed rail.

- Support the electrification of bus fleets.

- Let your local officials know that transit access is a priority, routes need to serve all communities, and transit stops need to have accessible sidewalks and safe places to wait.

Drive more efficiently

Unfortunately, not every community has mass transit infrastructure. And sometimes driving is the only practical way to get where you need to go, especially if you're moving many people, bulky cargo, or animals.

Nationally, 17% of carbon emissions in 2017 came from passenger cars and light-duty vehicles. In my home state of Washington, where energy production is cleaner, private transportation is the largest source of emissions. Clearly, this is an area where individual action can make a big difference. When possible, find ways to walk, bike, or use transit. But when the only practical option is to drive, do it as efficiently as possible:

- **Drive an electric vehicle.**

- Drive a hybrid.

- Carpool when it's feasible and safe to do so.

- Group errands.

- Optimize fuel efficiency (see sidebar).

- If you must use a rideshare or taxi service, use the carpooling feature or share the ride.

Go further

- Push for a clean electrical grid at the local, state, and federal levels.

- Support efforts to develop a robust network of accessible public charging stations.

Optimize fuel efficiency when you drive

- Avoid aggressive driving. Rapid acceleration and hard braking can increase fuel consumption by 15-40%.

- Keep your speed steady, at or below the speed limit. Allow your car to slow naturally uphill and speed up again downhill. To improve gas mileage 10-15%, drive 55 mph, not 65 mph.

- Avoid idling your vehicle when you're waiting for a ferry, train, or passenger. If you're stopped for more than 30 seconds—except in traffic—turn off the engine.

- Properly inflate your tires. Under-inflated tires can increase fuel consumption by as much as 6%.

- Cool the car strategically. Use the air conditioner sparingly in older cars. Newer cars have more efficient air conditioning, so using the AC may save more fuel than opening windows.

- Use cruise control strategically. It saves fuel by maintaining a steady speed in most conditions, but turn it off in hilly areas, where a steady speed requires extra acceleration and braking.

- Service your vehicle regularly. A poorly tuned engine produces up to 50% more emissions than one that is running properly.

If you drive a hybrid, also do the following:

- Use your brakes rather than coasting to a stop. Apply steady, even pressure. Braking recharges the battery.

- Maintain a steady speed under 55 mph. Accelerate gently to keep your car in EV mode.

- Use the car's idle-stop system. Putting your car in neutral prevents electrical recharging. Let your system automatically shut down the engine when the car has stopped.

Fly as a last resort

Flying is the fastest way to travel long distances, but it's also a hassle, physically uncomfortable for many of us, harms communities near airports, and generates huge emissions.

Traveling longer distances by train, bus, bike, and even a personal car is much more climate-friendly—and usually more pleasant, too. Instead of prioritizing speed, find ways to make the journey part of the adventure rather than something to endure.

Just how bad is flying for the climate? Not only do airplanes use jet fuel, but emissions at high altitude are three to five times worse than emissions at ground level. Beyond carbon dioxide, airplanes emit particles, nitrogen oxides, and sulfates that trap heat and have a much more damaging effect at cruising altitude.

One roundtrip flight from New York to Europe or California creates a warming effect equivalent to 2-3 tons of carbon dioxide per passenger. Scientists say we need to emit less than 3 tons of carbon dioxide per person per year, which means you can easily spend your entire personal annual carbon budget on a single roundtrip flight.

Unfortunately, unlike other areas of carbon emissions, air travel isn't likely to become more climate-friendly any time soon, even with tremendous political will. There aren't eco-friendly, effective alternatives to jet fuel yet, and there's no way to reduce the climate effects of flying without them.

Many climate activists and scientists grapple with the knowledge that flying is so devastating for the climate while also recognizing the need to visit the people we love, and that traveling makes us better world citizens. One compromise is to reserve flying for what many describe as "love miles," the flights that help you strengthen relationships, and for once-in-a-lifetime opportunities.

Before you book your next flight, consider alternatives:

- **Take a bus or train instead.** Especially for trips that are under 300 miles, a bus or train may even be faster once you consider security lines.

- Vacation closer to home. Travel by train, bus, car, or bike and explore the wonders of cities, mountains, lakes, beaches, or forests nearby. Have you enjoyed all that your region has to offer?

- Schedule a teleconference instead.

When you need to fly, reduce the impact:

- **Consolidate trips.** Make the most of the trip to get the greatest value from the carbon you've spent. Use ground transportation such as trains, buses, cars, and bikes for regional travel at your destination.

- Fly direct. Airplanes use much more fuel during takeoff than during the flight itself. Choose direct flights whenever possible to reduce the number of takeoffs.

- Invest in climate solutions. It's better not to generate the carbon at all, but if you must fly, invest in climate solutions to mitigate the effects of your flight. See "Reconsider carbon offsets" in this chapter.

Go further

- Let Congress, the Environmental Protection Agency (EPA), and the Federal Aviation Administration (FAA) know that you expect the U.S. to require cleaner jet fuels and more efficient airplanes.

- Support high-speed rail and investments in reliable public transportation options at the city, regional, state, and federal levels.

- Let airlines and the cities you fly to know that you expect direct flights with efficient bus and train options to get you the rest of the way.

Reconsider carbon offsets

Carbon offsets are a fashionable way to feel less guilty while we continue to prioritize convenience and habit over sustainable action. But they're not the panacea some would have you believe they are.

A carbon offset is an investment in a project designed to reduce greenhouse gases (solar farms, tree planting, etc.), meant to compensate for emissions made elsewhere. That is, the effect of your flying is supposedly not harmful because somewhere on the planet, someone is planting some trees that wouldn't have been planted otherwise.

Unfortunately, most so-called carbon offsets invest in projects that would have happened anyway. And sometimes offset programs result in indigenous people losing access to their traditional lands and natural resources.

The best option is to not generate pollution in the first place. Find alternatives to flying, skip the cruise, eat less meat, and generally be conscientious. But if you can't avoid a flight or other carbon-intensive activity, attempting to offset the damage is better than not doing anything.

According to Giving Green, three offset programs are worthwhile: Climeworks, Tradewater, and BURN.

Potentially more effective and more personally meaningful than purchasing carbon offsets is to make a sizable donation of time or money to an organization that is working to change climate policy, invest in renewable energy, or mitigate the effects of climate change on the communities most harmed.

I think Dr. Leah Stokes said it best: "My offset plan is activism."

Skip the cruise

They're marketed as floating paradise, but cruise ships do great environmental harm. One ship can emit pollution equal to 700 trucks and as much particulate matter as a million cars. Each passenger emits three times the carbon they would on land.

The ships release huge quantities of carbon dioxide, as well as sulfur, fine dust, heavy metals, and other particulates. The International Maritime Organization required all vessels to switch to cleaner fuel with a lower sulfur content by 2020. However, efforts to clean fuel have mostly been through scrubbers, which remove pollutants from the fuel but just transfer the pollution to the water.

Liquefied natural gas has been touted as safer, but it's still climate-hostile because it's obtained by fracking. Natural gas is a fossil fuel, and we simply need to leave those in the ground.

Cruise ship companies have been caught discharging sewage, toxic chemicals, plastic, and other waste into the ocean. The engine noise harms marine life; and hull paint sheds heavy metals into the ocean. Work conditions on cruises are abysmal, with crews working long hours for little pay.

Go further

- Let cruise lines know you aren't traveling with them until they have zero emissions, and meet other environmental, health, and labor standards.

- Share Friends of the Earth's cruise line report card with friends and family who take cruises: https://foe.org/cruise-report-card/

- Tell your representatives to address cruise ship pollution with regulations, research, and enforcement of existing laws.

Eating

For many of us, cooking and eating are among life's greatest pleasures. We can eat well and have a healthy climate, but it will require us to be a little more thoughtful about our diets. Food production accounts for about 25% of greenhouse gas emissions, so we clearly need to make some changes.

The technically easiest but emotionally most difficult action is to change *what* you eat. Common-sense and money-saving actions to prevent food waste are easier psychologically but still require being more intentional about shopping, food preparation, and food storage.

Every time you eat a meal or even a snack, you have a chance to move a little closer to the goals you set for yourself. Learning new ways to prepare meals can be a fun adventure, as can discovering many tasty cuisines that are based around sustainable diets. With something as emotionally charged as diet, you'll probably be most successful either starting small and building on your efforts or teaming up with friends or family members to challenge yourselves.

Eating responsibly also means considering the people producing our food. As more of our food is grown by larger corporations, local farmers need our support. And the people who plant, care for, and harvest crops deserve safe, fair working conditions. Supporting local farmers and farmworkers is an important part of creating a healthier, more sustainable food system.

Choose plant-based proteins

Hamburgers or black beans for dinner? Producing beef emits 20 times the emissions as producing beans, per gram of protein. Factory farms, which produce most of the meat available in the United States, not only release harmful emissions but create waste that pollutes our water, land, and air. They also create unhealthy, stressful, and sometimes deadly conditions for workers.

To meet climate targets, we must shift our diets from meat toward plant-based foods. Even if you're not prepared to adopt a vegan diet, you can probably find meat-free meals that delight you.

- **Eat little or no meat.** Especially avoid meat from ruminants such as cows, goats, and sheep (lambs).

- Choose proteins with as little processing as possible, such as tofu, beans, tempeh, seitan, and nuts. Highly processed meat substitutes can be less nutritious and more carbon-intensive.

- When you do eat meat, choose cuts from animals who have been raised responsibly. For a helpful guide, see Impactful.ninja.

- Start the transition by participating in Meat-free Mondays, or by planning one meal a day that is meat-free.

. . . and go easy on the dairy

Unfortunately, adding dairy products to everything in our diets isn't just affecting our waistlines or creating gastrointestinal issues. All that dairy is accelerating climate change, too.

Cheese, especially, is significantly more greenhouse gas-heavy than turkey, chicken, tuna, or eggs. The more milk it takes to make cheese, the more dairy cows produce methane and manure, pollute waterways, and consume resources.

If you really want cheese, lower-fat choices have a lower carbon footprint, so choose mozzarella, cottage cheese, feta, muenster, or ricotta.

Likewise, soft cheeses require less milk, so consider Brie, Camembert, and goat cheese rather than Parmesan and other hard cheeses.

- **Eat little or no dairy, especially cheese.** Lower-fat and softer cheeses are better than hard, aged cheeses.

- Use cheese sparingly as flavor enhancement instead of relying on its texture or flavor to carry the meal.

- Try alternatives to milk from cows or other ruminants. Soymilk, rice milk, and milk from various nuts can be tasty, nutritious substitutes. Check labels for protein content and added sugars.

Go further

- Insist that cafeterias in schools, hospitals, and other institutions in your community provide meat-free, dairy-free, nutritious, flavorful options.

- Support incentives for ranchers to graze cattle in a way that promotes soil health and carbon retention. Studies have shown that raising cattle can actually result in a net carbon decrease when cattle rotate infrequently through thoughtfully grown pasture.

Waste not, want not

Do you know what's at the back of your fridge? When you face unidentifiable leftovers, do they go in the trash or the compost? Landfills account for more than a third of our total methane emissions, and a large percentage of what ends up in landfills is wasted food.

Uneaten food is a problem on the other side of the equation, too. Nearly a quarter of greenhouse gas emissions come from food production and the changes in land use that accompany it. Excess food production increases emissions, worsening climate change, which in turn makes it more challenging to grow food—and more ecologically hazardous.

Wasting food means wasting water too. Fully one-quarter of US water consumption is used to produce food that isn't eaten.

Of course, wasted food is also wasted money. The average U.S. household throws away more than $1300 worth of food each year.

With some pretty minor lifestyle shifts, you can save money, preserve water resources, and help prevent food from ending up in a landfill!

- **Buy and cook only what you and your household will eat.**

- Plan meals, and stick to your shopping list. Need help getting started? Check out savethefood.com/meal-prep-mate/.

- Store leftovers and be creative using them to make tasty meals.

- Label containers with the dates you opened them or stored leftovers in them. Keep food that will expire soonest toward the front of the refrigerator or pantry so you'll see them and eat them.

- Compost food scraps in your home composter, your city's yard waste collection, or a community garden. You can even compost using a small box in an apartment or condo.

- Support local, sustainable farmers at your neighborhood farmer's market, food coops, through community-supported agriculture (CSA) subscriptions, and in your local supermarket.

Go further

- Support city and county composting programs.

- Support programs that let supermarkets donate food they can't sell to food banks.

- Support legislation that protects farmworkers' rights and creates less hazardous, less polluting conditions in food production.

Eat out responsibly

When you order meals for in-person dining or for takeout, support restaurants that follow sustainable practices in sourcing and cooking their food, providing meat-free and dairy-free options, composting, and offering compostable or recyclable take out containers.

- **Order only what you and your household will eat.**

- Take your own reusable containers for carrying leftovers home when you dine in at restaurants.

- Opt for meat-free meals, or seek sustainably sourced meat.

- Walk, bike, or use transit to pick up takeout orders.

Go further

- Let local restaurants and national food chains know you want meat- and dairy-free meal options.

- Support programs that let restaurants provide unneeded food to food banks and shelters.

- Support programs that recycle restaurants' used cooking oils into fuels for vehicles or other uses.

- Encourage restaurants to use compostable takeout containers and utensils.

- Support restaurants that provide delivery by bike.

What's the deal with palm oil?

Palm oil is everywhere: it's in half of all supermarket products, from frozen pizzas to margarines; used in body creams, soaps, makeup, and candles; included in detergents; and grown for biofuels.

Palm oil plantations cover more than 27 million hectares of the Earth's surface. Forests and human settlements have been—and continue to be—destroyed to make room for palm oil plantations, which provide virtually no biodiversity. Huge tracts of rainforest are being bulldozed daily in Southeast Asia, Latin America, and Africa to create even more plantations, releasing vast amounts of carbon into the atmosphere.

Palm oil-based biofuels have three times the climate impact of traditional fossil fuels.

So, it's bad news, right? Many groups have called for boycotts of palm oil altogether. But more than six million people—mostly subsistence farmers—rely on palm oil for their livelihoods. Using alternative fats is potentially *more* damaging to the climate. And it's possible to grow palm oil without hurting the environment. Industry and environmental groups have worked together to create growing standards toward a climate- and people-friendly growing practice. NDPE standards require palm oil to be produced without deforestation, peat development, or exploitation.

As of 2020, the Roundtable on Sustainable Palm Oil (RSPO) certification system certified 20% of the world's palm oil supply.

So with all that in mind, avoid palm oil when possible. To ensure the brands you buy are using only responsibly grown palm oil, check out the World Wildlife Foundation's palm oil buyers scorecard at palmoilscorecard.panda.org.

At home

Electricity and natural gas are so integrated into our lives that we don't even notice we're using them until the power goes out or the gas is turned off, and suddenly we can't do all the things we normally do. All that energy use is contributing to global warming. The less we use, and the cleaner the sources of that energy, the lower our greenhouse gas emissions.

Generally, you use the most energy to make something hotter or colder, so you can cut your emissions fastest by focusing on more efficiently heating and cooling your living space, your food, your water, and your clothes.

Just as important, ensure that the energy you use is clean. Remove fossil fuels from your home environment (including coal, oil, propane, and natural gas) to not only reduce emissions that contribute to global warming but to clean up the air you're breathing in your home! To gain access to cleaner energy, push your utility to phase out fossil fuels and ramp up renewable energy production.

Conserving energy is important even if your local grid is clean. Excess renewable energy can be exported to communities that rely more heavily on fossil fuels.

Many of the changes that make a difference cost money, and if you rent, you may not control the energy efficiency of your home. So it's also important to support policy and financial initiatives to make it easier and less expensive for everyone (and more advantageous for landlords) to make meaningful changes.

Heat and cool efficiently

Generally, you use the highest intensity of energy when you heat something up or cool something down. The biggest thing you heat and cool is the space you live in. Anything you can do to reduce the energy you use to keep your space comfortable is likely to reduce emissions and save you money in utility bills.

- **Use efficient heaters and air conditioners.** Mini-split ductless heat pumps are cost-effective and climate-friendly. Radiant heat, active solar heating, and geothermal heat pumps can also be great solutions. Even upgrading your current boiler or furnace to an energy-efficient model can save significant energy. Many utilities offer rebates and guidance for installing energy-efficient heating and cooling appliances.

- Lower the thermostat in winter; raise it in summer. The U.S. Department of Energy suggests setting your thermostat to 68° in the winter and 78° in the summer. Smart thermostats can be programmed to automatically lower the heat or AC when you're sleeping or away from home.

- Weatherize your home, and use insulating blinds and curtains. Energy-efficient windows can help keep heat in your home in the winter and out in the summer.

Cook cleanly

Many notable chefs are touting the superiority of electric induction cooktops over polluting, wasteful gas ranges. For those of us who are just trying to get dinner on the table, a less-expensive electric range is usually fine.

- **Replace natural gas ranges with electric.**

- Use the least energy-intensive methods to cook. Ovens and stove-tops require the most energy; microwave ovens, electric kettles, and toaster ovens require less. In the summer or in especially sunny areas, use an easy-to-build solar oven!

- Keep the freezer and fridge full for more efficient cooling. Crumpled newspaper is an inexpensive and effective way to fill in gaps in the freezer.

Go further

- Urge your utility to provide incentives for people to switch from natural gas appliances to electric.

- Support efforts to ban natural gas in new buildings and transition existing buildings to become all-electric.

Use less energy throughout your home

Especially if your local grid is powered by coal or other fossil fuels, energy conservation is key. Remember that you'll have the greatest gains if you focus on heating and cooling.

- **Buy energy-efficient appliances.** For helpful information, see www.energy.gov/energysaver/appliances-and-electronics.

- Use a home energy monitor to identify what's sucking down the most energy, so you can make changes where they're most useful.

- Hang clothes outside to dry, or inside if the weather is bad or you don't have access to appropriate outdoor space. In the US, clothes dryers represent 4% of an average home's total energy use.

- Reduce the energy used to heat water: Take shorter showers, wash clothes in cold water whenever possible, and lower the temperature of your water heater when you're away for more than a few days. Consider installing an on-demand water heater, which uses energy to heat water only when you need it.

- Use the dishwasher instead of hand-washing dishes, and run it only when it's full. Turn off heated dry in the dishwasher and let your dishes air dry.

- Run only full loads in the clothes washer.

- Turn off lights and gadgets when you're not using them. Use smart power strips to eliminate standby power use (also called "vampire energy").

- Use hand tools—brooms, manual lawnmowers, and screwdrivers—instead of power tools when possible.

- Change light bulbs to LED.

Go further

- Support local, state, and federal initiatives to weatherize homes and replace appliances for low-income households.

- Advocate for incentives and penalties that encourage property owners to make their tenants' homes more energy efficient.

- Where marijuana is legal, support efforts for growing it outside on farms, rather than in indoor grow operations, which require intense lighting. In 2012, about 1% of all electricity used in the U.S. went to grow operations!

Switch to clean energy

Renewable energy sources (solar, wind, geothermal, hydro) are cleanest. Coal and oil are arguably the dirtiest, and it's become clear that natural gas is just as bad. Whenever possible, you're better off ditching fossil fuels and choosing cleaner sources, either on your property or through your city utility.

- **Replace natural gas appliances.** Use electric water heaters, furnaces, clothes dryers, and cooking ranges. Natural gas pipelines leak methane and sometimes explode. Fracking damages communities, pollutes water supplies, causes seismic activity, and leaks heat-trapping methane. Natural gas has gotten a pass for too long—it's a fossil fuel, and we need to leave it in the ground.

- If you own your home, install solar panels. Every year, solar panels get more accessible, affordable, and efficient. Typically, home solar energy production is connected to the grid and you may be able to make money selling clean energy back to your local utility. If you want to have electricity during power outages or just want to be off-grid, in most cases you can attach a battery to store the solar energy you produce.

- Invest in community solar projects. In many communities, you can chip in for solar installations at public buildings such as community centers or schools and receive credit against your utility bill as those panels generate power.

Go further

- Tell your local utility that you want renewable energy.

- Find out whether your state has renewable energy requirements. If it doesn't, let your state legislature know that you support them. (If you don't think they're high enough, let them know that, too!)

- Support efforts to make renewable energy, efficient appliances, and home weatherization more accessible and affordable through community-buying programs.

- Support incentives to install renewable energy, including rebates and the ability to sell energy to the grid.

- Tell legislators that "clean energy" sources required by legislation must be truly clean. Natural gas is not a bridge fuel. Hydrogen sounds appealing because it's so widely available, but it's not clean when you mix it with natural gas; the nitrogen oxide that is produced is polluting vulnerable communities.

Garden sustainably

In the United States, individual yards provide 42 millions acres of land that could potentially capture and store carbon. How you care for your yard, whether it's a grassy lawn or a vegetable garden, can help or hurt the effort to curb climate change.

- **Plant hardy perennials.** Native plants that don't require water once established can be much more attractive and easy to care for than thirsty grass lawns. Tour local native gardens to see ideas for healthy lawn replacements in your region.

- Grow your own organic food! For inspiration and tips, see greenamerica.org/climate-victory-gardening-101.

- Limit your use of nitrogen fertilizer. Excess nitrogen is consumed by bacteria that converts it into a greenhouse gas 300 times worse than carbon dioxide, and homeowners pollute at a rate 10 times greater than farmers. Nitrogen also finds its way into waterways, creating oxygen-starved dead zones in oceans and lakes.

- If you have grass, use a manual push mower instead of a gas or electric mower. You'll get a better workout, and with a sharp blade, it does the job just fine. Then leave the grass clippings to mulch.

- Use a broom or rake instead of a leaf blower, either gas or electric. Manual labor doesn't emit carbon, and it's also more efficient, quieter, less polluting, and much easier on your back.

- Plant trees. With their deep roots and large biomass, trees store carbon efficiently as they provide oxygen, shade, habitat, and food for birds and other species.

- Consider planting bamboo, which sequesters carbon faster than almost any other plant. Just make sure to use precautions to sequester the bamboo as well as the carbon; bamboo can be invasive and difficult to eradicate once established.

Renewable natural gas

Methane is an incredibly harmful greenhouse gas, trapping 30 times the heat of carbon dioxide. We hear a lot about methane from cows, but a large percentage of methane emissions come from wastewater treatment plants and landfills. Capturing that methane and converting it into a gas diverts the methane from the environment and provides usable energy. That's a good thing, and where methane exists (waste management, industry, and agriculture), we should encourage the use of anaerobic digesters and other technologies to convert the gas where good alternative energy sources don't exist.

In fact, California's climate goals include capturing 40 percent of methane by 2030, converting it to biogas.

However, renewable natural gas is not a panacea. It's greenest if it's used onsite, powering the farm or utility that produces it. When it's piped away, it encounters the same leakage problems that plague fracked gas. And even if all existing methane were converted to biogas, there's not enough to fill the niche that fracked gas currently serves. Gas infrastructure is also hazardous, and it requires major safety upgrades and expensive investments in leak mitigation.

We should capture as much methane as we can, but meanwhile, we should be working to eliminate methane emissions at landfills, reduce the emissions at dairy farms, and disassemble the natural gas infrastructure as we transition to truly renewable, sustainable energy sources.

Biomass

Biomass is any living organic matter that's used for fuel. There's something comforting about a wood fire, but that comfort fades when you see the carbon emissions behind the flames—and that's especially true when you move beyond an individual fireplace to a massive incinerator.

Forty-four percent of bioenergy in the United States comes from wood, cut from forests and burnt in biomass incinerators that emit higher levels of carbon dioxide than most coal-fired plants. Some wood pellets are created from sawdust or unusable waste from industries, but increasingly, they come from clear-cut forests. Deforestation and forest degradation together are a huge source of carbon emissions, second only to the burning of fossil fuels.

What about other forms of biomass? The net effect of biomass depends on its source, growing conditions, what would otherwise be grown, and the process by which it's converted to energy.

Historically, biofuels have been produced from food crops such as corn and soy. Ethanol, for example, is a gasoline substitute derived from corn that is politically popular in the early-primary state of Iowa. The theory is that because corn pulls carbon dioxide out of the air as it grows, the carbon dioxide released when we burn it is the same carbon it absorbed, resulting in net-zero emissions. But when you consider fertilizer, farm equipment, transportation, energy used in the conversion to liquid fuels, and potential deforestation, it's not such a sweet deal.

Some biofuels may play a useful role, however. Clean biofuels come from crops that don't require much fertilizer, water, or other care. Switchgrass has potential, provided it's grown on farmland unsuitable for most food crops, especially if grown near a refinery.

- Compost your food scraps and garden waste. Add leaves to your compost pile or garden beds. If you don't have a use for them, offer them to neighbors. (Leaves are far too valuable to send to the landfill!)

- If you're creating a patio, driveway, or other hard surface, consider alternatives to concrete. The chemical process used to create cement (a component of concrete) emits high levels of carbon dioxide, largely because the high temperatures require burning coal. Natural stone, wood, or recycled composites may give you the surface you want without the damage to the climate.

Go further

- Advocate for your city or county to collect yard waste and food scraps to compost.

- Support efforts to plant, maintain, and preserve trees.

- Support the development of community gardens, including roof gardens.

- Become a Master Composter or Master Gardener and teach others how to compost and garden effectively. Most states offer training in exchange for a volunteer commitment.

- Ask whether your parks department uses pesticides. If they do, urge them to transition to pesticide-free maintenance to support the pollinators, birds, insects, and other living creatures we need for healthy soil.

- Urge your state and federal representatives to support climate-friendly agricultural practices, such as reducing tillage, planting cover crops, rotating crops, integrating livestock and crop production, and amending soil with biochar.

Regenerative agriculture

If you manage larger areas of land, you may be able to not only curb emissions but actually draw carbon out of the air. Industrial agriculture has caused many problems for the climate—and significantly reduced the land's capacity to produce healthy crops. But transforming agricultural techniques can help with climate change, feed more people with more nutritious food, and provide farmers with profit.

A variety of science-based practices including composting, adding biochar, no-till farming, and pasture cropping restore topsoil and the vital healthy soil ecosystems that we rely on while also sequestering carbon. Some of the most effective practices were in use long before pesticides and heavy equipment were developed, and it's time to return to those practices and accelerate their adoption.

To learn more and help spread the word, check out The Carbon Underground (thecarbonunderground.org) and Regeneration International (regenerationinternational.org).

Kiss the Ground—the book and the documentary—provides a great introduction to the importance and the promise of regenerative agriculture.

Stuff

When my three-bedroom house was built in 1888, there was a single closet at the end of a hallway. Partly that was because people tended to use armoires and other furniture to store things, but mostly people just didn't have that much stuff.

Now we have so much stuff that offsite storage has become a big industry. We don't use most of the stuff we own, and we throw things away at the first sign of wear. But everything we acquire took energy and raw or recycled materials to make, and everything that goes to landfills represents wasted resources.

Marie Kondo has the right idea: we should only have things that we value, whether it's because they serve a practical function or they bring us joy. But the key isn't just to get rid of what you don't want, it's to acquire more judiciously in the first place.

Be a conscientious consumer. Think about the lifecycle of the things you purchase—the resources that went into creating them, the people who created them, the packing materials used to ship them, the emissions spent transporting them, your use and enjoyment of them, and their ability to be reused or recycled when you're done. Making good decisions about every part of that lifecycle can not only reduce emissions, preserve resources, and improve working conditions, but also clear space in our lives so that we can truly enjoy what we have.

Consume strategically

We're told that shopping is recreation; the things we own have become our way of expressing ourselves. But it doesn't have to be that way. You can save money and enjoy life more by putting the credit card away, appreciating the things you already have, and participating in experiences: play a board game with the kids, sing with friends, hike through nearby woods, picnic in the park, play fetch with the dog, or snuggle up with a library book.

All the stuff we buy, use, and discard requires massive amounts of energy and natural resources, contributing to global warming through energy use on the front end while creating pollution, choking sea life, and contributing to methane-producing landfills on the back end. The less we consume, the smaller our carbon footprint. To learn more about how our consumption habits affect the environment, workers, and other species, check out storyofstuff.org.

It's not all up to us as consumers, either. Smart regulations, including producer responsibility programs, give manufacturers strong incentives to find more sustainable ways to produce goods and package them.

- **Consider whether you need the new thing you're about to purchase.** What could you do if you don't buy it?

- Research before you buy. Find the most efficient, least resource-intensive version of whatever you need, whether it's a refrigerator or a new pair of running shoes.

- Buy quality. Things that last won't end up in the landfill or need to be recycled as soon.

- Use, maintain, and repair your cellphone, tablet, computer, television, and other electronic devices for many years. These devices require rare elements, often mined in hazardous and exploitative conditions, and most result in toxic pollution when you're done with them, sometimes even when you recycle them responsibly. Chances are good an older model will serve you well for several years.

- Join your local Buy Nothing, Freecycle, or other neighborhood sharing group. Give the things you no longer need or want to people who can make good use of them, and delight in the things others no longer want. If a group doesn't exist in your area, start one!

- Try to repair things before tossing them.

- Buy fixtures and other home improvement materials from second-use stores. Often, items salvaged from older homes are of higher quality than newer building materials.

- Consider giving experiences, such as theater tickets or cooking lessons, rather than things to commemorate birthdays or holidays.

Go further

- Support local efforts to ban single-use plastics, where practical.

- Support producer responsibility laws, which make producers responsible for the life cycle of products that are hard to dispose of, including post-consumer collection, recycling, and/or disposal.

- Start or support community tool libraries, repair meetups, public libraries, and other local efforts to share resources.

Avoid unnecessary packaging

So much plastic and paper go into packaging that we discard immediately. Whenever feasible, choose products that use less packaging (or none!). Because individual choices aren't enough, let manufacturers and legislators know that we need to reduce packaging and that any packing materials that are required should be reusable or recyclable.

- **Carry a durable water bottle.** Tap water is often higher quality than bottled water—check with your local utility to learn about your water. The Tap app (findtap.com) is a free app that lets you find sources of free tap water to fill up your water bottle. Many public water fountains include convenient bottle filler stations.

- Buy food and spices in bulk, using glass jars or other sturdy, reusable containers.

- Use durable, washable produce bags to reduce the number of single-use plastic bags you use.

- When possible, carry a small set of cutlery, plate, bowl, and cloth napkin, especially if you're attending a potluck. Take your own reusable container to a restaurant for leftovers.

- Find creative alternatives to single-use packaging at zero-waste stores, which are available in most states. Search "zero waste store" online to find options near you.

Go further

- Let the stores you patronize know that you want less packaging, especially plastic.

Minimize shipping emissions

Buying goods online is incredibly convenient, but shipping to individual homes can be less efficient than shipping to local stores.

As with most things, there are nuances. It's more efficient for one delivery truck to travel a route to several homes than for each resident to drive to a store and back. But deliveries are much less efficient for those of us who get to the store on foot, by bike, or using transit.

There are many advantages to buying locally whenever possible. Smaller stores in your area need your business and typically provide better, more attentive service, but even chain stores provide paychecks to local folks. It's also less stressful to get what you need immediately.

You build connections with community when you patronize local businesses, too, as you see your neighbors and get to know the people who work there. In many small towns, the local Walmart is where neighbors catch up on Saturday mornings!

- **Walk, bike, or take transit to local brick-and-mortar shops.** You'll support local businesses and your local economy, reduce your climate emissions considerably, and get what you want immediately.

- Resist the urge to have packages shipped overnight unless the need is actually urgent. Usually, you can plan ahead and select ground shipping, which is much more environmentally friendly.

Go further

- Tell online retailers that you want ground shipping options.

- Support Small Business Saturdays and other events to promote small local businesses.

Be a smart recycler

Recycling can reduce the emission of carbon dioxide and methane by saving energy and natural resources in the creation of new material and by diverting usable materials from landfills. However, recycling is only beneficial when it's done appropriately and conscientiously. Materials that are not recyclable—either due to their size or content, or because they're not clean—can contaminate an entire load and condemn it to the landfill. In particular, be careful about plastic recycling—most of it isn't recyclable, no matter what the industry has been saying for years.

- Reduce the amount of packaging you bring into your home to begin with. You don't have to worry about how to discard something you never had.

- Follow the instructions you're given by your local utility closely. Toss into the recycling bin only the things that your utility accepts.

- Clean and dry plastic and glass containers before putting them in the recycling bin.

- Compost soiled paper and cardboard; don't try to recycle it.

- When in doubt, throw it out.

- Recycle electronics, batteries, CFL lightbulbs, and other things that require special handling at designated certified recycling centers. States and municipalities have different regulations; an online search for "electronics recycling" and your city or state should help you find an appropriate recycling center. For more information, see https://www.epa.gov/recycle/electronics-donation-and-recycling.

Not-so-fast fashion

One of the major sources of emissions is lurking in your closet—and on the shelves of stores and warehouses all over the world. With the advent of "fast fashion," clothing is less expensive and less durable, designed to be worn only a few times before the next trend arrives on the scene. There are approximately 20 new garments manufactured per person per year; we are buying 60% more than we were as recently as 2000. What's worse: if the trend continues, the clothing industry will triple its resource consumption between 2000 and 2050.

Textile production is one of the most polluting industries, producing more emissions than international flights and maritime shipping. In fact, the apparel and footwear industries combined contribute 8%-10% of global climate emissions.

Beyond climate emissions, toxic dyes pollute waterways; the production of wood-based fabrics such as rayon, modal, and viscose causes deforestation; polyester fabrics shed plastic microfibers into waterways; and cotton production uses large amounts of pesticides and especially water. The industry also exploits workers, who work long hours in unsafe conditions for little pay to make cheap clothing.

So what's the answer? Buy less, choose quality, and take care of it. If garments were worn twice as many times as they currently are, on average, greenhouse gas emissions from apparel would be 44% lower.

- **Buy used clothing.** You'll give garments a second life, and it's fun to hunt for treasures at thrift stores or consignment shops!

- Buy high-quality clothing and shoes, in styles you'll want to wear for a long time.

- Wash clothing infrequently; use cold water to save energy and make your clothes last longer.

- Hang-dry clothes whenever possible, both to conserve energy and to preserve clothing. Clothes dryers are hard on some fabrics.

- Donate used clothing and shoes to local thrift shops or clothing drives. If clothing is torn or stained, tear it up to use as rags or donate it to Goodwill for fabric recycling.

- Mend clothing yourself or hire a mending service.

- Do your homework before you buy new clothing. Buy clothes from brands that are producing durable clothing sustainably and treating their workers well.

- Shop locally to reduce the carbon emissions from shipping and packaging.

- Buy recycled polyester fleece, which uses less energy than original production.

- Participate in clothing exchanges if you like to try new styles or tire of wearing the same thing; these are especially useful if you have kids who grow out of clothing and shoes quickly. You can also use clothing rental services, but be aware of the emissions created by shipping garments back and forth.

Go further

- Support producer responsibility laws, which make producers responsible for the life cycle of products that are hard to dispose of, including post-consumer collection, recycling, and/or disposal.

Raising kids

If you're a parent of an infant or school-aged children, you probably spend a lot of your time and energy ensuring they have what they need to thrive. It's an incredibly important role. The generation you're raising will play a critical role in the future of our species and others on this planet.

Parenting has always been complicated, but it's even more so now when there are so many crises. From deciding whether to have children to raising your kids to be compassionate and climate-friendly citizens, the conscious choices you make can help limit the climate emissions your family produces while amplifying your child's positive effect on the world. As you make these choices for yourself, it's also important to ensure others have choices too. Support reproductive justice, including access to contraception, abortion, assistive reproduction technologies, and prenatal care.

Parent—or don't—intentionally

With climate change posing an existential threat, it's ironic that one of the most effective ways to reduce emissions is to choose *not* to perpetuate our species.

For the sake of the children who *are* born, every child should be wanted—and raised to minimize their impact on the climate. While the United States does not have the population density of some poorer countries, every person in this country—even those living in poverty—has a much larger climate footprint than people living in other parts of the world.

- **Plan pregnancies.** Heterosexual couples, use birth control when you're not actively trying to get pregnant—and guys, when you've had your kids, consider having a vasectomy.

- Be thoughtful when considering how many children to have. While there are many factors involved in such a personal decision, consider reproducing yourselves and no more. (That is, if there are two parents, have two children.)

Go further

- Support reproductive justice efforts so that everyone can plan their families. That means that women and their families get to choose whether and when to have children, and that they have the resources they need.

- Tell your senators and representative to provide federal funding for family planning.

- Tell your city, county, and state elected officials that you support family planning resources and access.

- Support education for women and girls internationally.

Evaluate diapering options

Disposable diapers are a huge source of plastic consumption, requiring fossil fuels and energy, and they end up in landfills, which produce methane. They're also incredibly expensive.

Cloth diapers are reusable, but require large amounts of water and energy to clean—and if you use a service, they can be just as expensive as disposables.

However you diaper your child, starting potty training early can be good for the child, good for the planet, and good for your wallet.

- **Consider Elimination Communication**, a process by which babies and their caregivers communicate so that babies can pee and poop directly into the toilet. You'll probably still need some diapers, but not as many. Learn about it at diaperfreebaby.org.

- Consider cloth diapers. Diaper services are available in most cities and cost about the same as buying disposable diapers. For a less expensive option—albeit with more hassle—you can make or buy and clean your own diapers.

- If you decide to use disposable diapers, choose a brand that uses recycled or renewable, nontoxic materials.

Prioritize quality

With babies come new furniture, clothes, toys, and gadgets. And as kids grow, they acquire new clothes, beds and desks and beanbag chairs, toys and books, bikes and skateboards and worm farms.

All of that stuff has an effect on emissions. You can reduce the toll on the climate, your budget, and your storage space by approaching purchases and advising gift-givers thoughtfully.

- **Opt for fewer toys.** Choose those that are durable, educational, and likely to continue entertaining a child for a long time.

- Repurpose mundane things that children might use as toys. Remember that little kids are often as happy playing with the box as with its contents.

- Choose durable clothing that can be altered as the child grows.

- As your child outgrows something, pass it on. Take on someone's else's hand-me-downs instead of buying new. (Bonus: you'll save money and build community, too.)

- Look for toys that don't require batteries or electricity.

- Join or create local toy and clothing exchanges with other parents.

Make responsible transportation fun

You don't have to get a minivan or SUV when you have kids. In fact, many families don't own vehicles at all. While it may be more challenging to herd young children on a bus or get them ready for a bike commute than it is to load them into a car, it's often much more cost-effective and the kids have experiences that they wouldn't if they were isolated from the outside world as they travel.

- **Walk to nearby grocery stores, parks, or school.**

- Bike to school, the library, or other destinations. There are supportive communities of family bikers in most areas of the country. They can help you think about what's doable for your family's circumstances, what kind of bike and other resources you'll want (hint: family bikers tend to carry snacks), and what routes are most comfortable. Family biking looks different with young children in bike seats or trailers than it does with bigger kids on their own bikes; kids often enjoy the ride either way.

- Take the bus or subway. Young children, especially, seem to love public transit. Things that adults take for granted are exciting to them. Let your child pull the cord to indicate you want the next stop, or read the route map together. Preteens and teenagers can learn how to navigate transit while riding with you so that they can feel more comfortable traveling independently when appropriate.

Go further

- Coordinate a walking school bus or bike train at your child's school.

- Carpool or bike with other families to school events, church, festivals, and other destinations you have in common.

- Let your city's department of transportation and city leaders know that you want safe routes to walk and bike for people of all ages, including young children.

Educate and learn from children

Around the world, young people are leading the charge on climate action. You can help your kids understand the role we all play in the environment around us—and then give them the support and resources they need to get involved.

- **Discuss as a family how to reduce your emissions.**

- Invite your kids to challenge all family members to choose climate-conscious practices over convenience—and to join with others to demand effective, equitable policy changes.

- Provide emotional support if your child is anxious about climate change. It's their future we've threatened, after all. Help them channel their anxiety or anger into action.

Go further

- Support school policies that allow absences for participation in climate and civil rights actions.

- Donate or provide other material support to youth-led climate action groups.

Money

Anyone invested in political activism knows how important it is to follow the money to see who's funding campaigns or other efforts. Just as important is to ensure that your money follows your values.

Even small sums can make a difference, depending on where you bank, invest, and donate. You also probably have influence over retirement or investment accounts at your workplace, college, or city.

As they say, money talks. So make sure yours represents you well.

Put your money where your climate goals are

While you're intentionally reducing your greenhouse gas emissions, make sure your money isn't working at cross-purposes. Whether you're a high-earner or living paycheck-to-paycheck, financial institutions use the money you deposit—or the interest you're paying on loans or credit cards—to fund other projects, including longterm investments in fossil fuel extraction and infrastructure. Without that funding, those projects can't go forward.

- **Use a climate-friendly bank.** To find an institution that offers the services you need and doesn't fund fossil fuels, visit bankforgood.org or mightydeposits.com.

- Join a local credit union. Credit unions are not-for-profit community-based financial institutions.

- When you move your account or cancel a credit card, let the bank know that you're doing it because they fund fossil fuel projects.

Go further

- Join efforts to pressure banks to stop funding fossil fuel projects. The Stop the Money Pipeline coalition (stopthemoneypipeline.com) provides ways that you can get involved.

Invest in the future

Do you have an IRA, 401(K), pension contributions, or any other investments in mutual funds or individual stocks? You can ensure that your money is not going to climate-damaging companies. And you can invest in clean technologies such as solar, wind, and geothermal power projects. As a shareholder, you may also be able to change the practices and policies of companies you're invested in.

- **Invest in socially responsible funds.** Choose funds that screen out investments in fossil fuel projects. As You Sow (asyousow.org) provides tools to help you assess specific funds.

- If no socially responsible funds are available through your employer's retirement plan, let them know that you want better options.

- Find out where your pension fund is invested. If the fund includes investment in fossil fuel companies, petition the fund managers to divest from them.

- Actively invest in renewable energy infrastructure.

- Join with other shareholders to propose and pass resolutions to address company policies and environmental practices. Learn about effective shareholder action from As You Sow (asyousow.org), which promotes environmental and social corporate responsibility through shareholder advocacy and innovative legal strategies.

Go further

- Let your representatives know that you expect the federal government, including the SEC and Department of Labor, to define and regulate ESG funds, which are funds that consider environmental, social, and governance factors.

Fund the movement

Philanthropy can also be a powerful tool to advance equitable, effective climate action. There are many fantastic groups organizing to pass effective climate legislation, hold polluters accountable, ensure equity in climate solutions, provide assistance to those most affected by climate change, educate women and girls, change policies at all levels, and challenge corporate power. No matter what you're most passionate about or how much money you can spare, there's almost certainly a group that could make great use of your donation.

- **Pick an area (or three) that you're passionate about.** Give to organizations that are doing effective work in that area.

- In particular, support groups led by the populations most vulnerable to pollution and climate change: low-income, Black, indigenous, people-of-color, disabled, and LGBTQ communities.

- If you are hoping for a tax deduction, contribute to groups that focus on education, providing service, and building community. Most groups' donation forms—printed or online—let you know whether the donation is tax-deductible, but if you're not sure, ask!

- Visit ClimatePocketGuide.com for a list of international, national, and regional organizations that could use your support.

Cryptocurrencies? Maybe. Bitcoin? That's a big no.

Bitcoin may seem sexy, but it's a huge problem for the climate. A single Bitcoin transaction expends enough electricity to power the average US home for three weeks—and that energy is coming mostly from fossil fuels.

Bitcoin is "mined" using computers to solve increasingly difficult problems. It's that mining process that sucks down so much electricity.

To be clear, blockchain technology is not inherently energy-intensive, and there are some amazing uses for blockchains. It's the proof-of-work verification method that Bitcoin uses that is the problem. Other cryptocurrencies use proof-of-stake verification methods that do not require obscene amounts of energy.

This can all be pretty complicated for anyone not deeply tech-proficient. But what most of us need to know is that Bitcoin is making a few people rich, more people greedy, and all of us more vulnerable to a warming world.

Changing institutions

While you're reducing your emissions, look beyond your own home and your family and friends. One person avoiding single-use plastics is helpful, but when institutions change purchasing habits, it's much more impactful. Your deciding not to fly for vacation is meaningful; your entire company rethinking its habit of flying employees across the country for meetings can have a much more dramatic effect.

Think about the organizations and institutions that are part of your life, especially those where you have the greatest influence: your school, your workplace, your place of worship, community organizations, and local governments. Working with a few like-minded people, you can change policies and behaviors—and set an example for other institutions and governments as well.

Collective action to change institutions can not only curb greenhouse gas emissions but also build community, promote equity, inspire other groups, and empower individuals.

At school

School communities provide great opportunities for action. They're typically smaller communities of people who share a common purpose and values.

Whether you're involved with an elementary or secondary school, community college or university, private or public institution, you can influence the school's policies and actions—and provide an example for other schools to follow. Students, parents, staff, alumni, donors, and community partners all have clout.

Depending on the school and your own experience, you might start with transportation, energy use, cafeteria practices, investments, political action, or curriculum. It's often most effective to start with a single, tangible accomplishment, and then build on that with the allies you develop along the way.

Transportation

When school is in session, students, faculty, staff, and visitors all travel to and from school buildings every day. Additionally, teams and supporters travel to competitions, and schools send faculty, staff, and students to conferences and meetings.

Elementary/middle/high schools

- **Coordinate a walking school bus or bike train** to help students get to school safely with a low carbon footprint. More students walking or biking to school also means less traffic at school drop-off and pick-up times.

- Perform a safe routes to school audit and advocate for changes, such as crosswalks, pedestrian signals, or sidewalks, that will make it safer for students to walk to school.

- Encourage a bike club, advocate for secure bike parking, and conduct a bike-maintenance workshop at a local middle or high school.

- Teach bike and pedestrian safety classes as part of the physical education curriculum. Local bike advocacy groups or your local department of transportation may be able to help with the curriculum and access to bikes for students during class.

- Support bike-to-school day and walk-to-school day events.

- Advocate for electric school buses, especially where the energy grid is clean.

- Educate parents not to idle vehicles at student pick-up.

Colleges/universities

- **Provide incentives to students, faculty, and staff to walk, bike, or use transit to get to school**: offer transit passes, bike facilities, and financial incentives. Conversely, charge for parking private motor vehicles. Start a bikeshare program on campus.

- Make sure shuttles are electric, especially where the grid is clean.

- Audit the campus for pedestrian safety and convenience. Are building entrances designed for people walking or for people driving? If streets run through campus, do they have slow speed limits, good sightlines, and safe crossings?

- Install pedestrian lighting, emergency call boxes, or other amenities students and staff require to feel safe walking at night.

Energy

In many school districts, energy costs are second only to salaries. Conserving energy reduces the school's greenhouse gas emissions and saves money that can be better spent to educate students. For inspiration, look to Batesville, Arkansas, where a large solar installation paired with new lights, heating and cooling systems, and windows helped the district move from a $250K budget deficit to a $1.8 million surplus.

- **Ensure that buildings are heated and cooled efficiently.** Update and maintain HVAC systems, insulate buildings sufficiently, and don't heat or cool rooms that aren't in use.

- Install solar panels, wind turbines, or geothermal heat pumps.

- Use efficient LED lighting. Install motion detectors in rooms that are used intermittently (such as restrooms), and make sure staff and students turn off lights in rooms that aren't used.

- Turn off computers, projectors, and other equipment when not in use, especially on Friday afternoons or before holidays. Use smart power strips to eliminate energy vampires.

- Take advantage of natural daylight and natural air circulation whenever possible.

- Use energy efficient audio-visual equipment and devices.

- Upgrade kitchen appliances, from refrigerators and ranges to microwaves.

- Turn off the lights in vending machines. (This seems minor, but Seattle Public Schools saved $20,000 a year by doing this.)

Cafeteria

A school community can do a lot to promote health, wellness, and responsibility while reducing waste.

- **Provide nutritious, appetizing foods that students will eat.** Invite students to help choose and plan the menus.

- Reduce or eliminate the use of single-use utensils, dishes, and packaging. Use washable, durable trays, not disposable trays or dishes.

- Install water coolers with washable glasses instead of providing individual plastic water bottles.

- Reduce food waste with a sharing table, or collect uneaten food to distribute to students whose families can benefit from extra food. Where students don't pay for individual items, start a campaign to encourage students to take only what they'll eat.

- Engage students in climate-friendly efforts such as Meatless Mondays, a Cafeteria Ranger program for zero waste, or planting a student garden to grow vegetables for the cafeteria.

- Ensure there are good vegetarian and vegan options available at every meal. De-emphasize meat and cheese in meals.

Investments

School employees usually have pension plans; colleges and universities have endowments and other investments. Fund managers answer to donors and community members, so you can influence them to shift investments away from fossil fuels and to renewables and other sustainable measures.

- **As an alumnus, advocate for your alma mater to divest from fossil fuels and invest in renewables.**

- If you're a student, organize teach-ins or campaigns to influence fund managers.

- Advocate for public school pension investment policies that include community values as well as financial return in the criteria used by fund managers.

Political action

Students are leading protests and community action on climate change all over the world. Other members of the school community can support their efforts.

- **Encourage youth to lead**, and provide them the support—including financial resources—that they need.
- Encourage students to join #FridaysForFuture strikes in front of their nearest city hall.
- Urge your college or university to include climate action in its agenda when lobbying state legislatures or city and county councils.
- Register eligible students to vote.
- Ensure your school has a robust climate action plan.
- Support youth lobbying days and other youth actions.

Curriculum

Informed citizens are better able to take meaningful action in our rapidly changing world.

- **Demand that elementary, middle, and high schools teach about climate change.** Make sure your school board requires an accurate, age-appropriate climate change—and climate action—curriculum.
- Urge your college to include classes in climate science.
- Support student groups focused on climate action.

At work

Before the pandemic, most of us spent more waking hours at work than we did at home. Curbing climate emissions at our workplaces can have more dramatic effects than individual action at home. Choose an area to focus on, and build on it with other colleagues who recognize the need for action.

Transportation

How employees commute to work and travel for work makes a big difference in climate emissions.

- **Charge for parking** to encourage employees and guests not to drive alone.

- Provide financial incentives for walking, biking, transit, and carpooling—even kayaking, if your location supports it!

- Make it easier to use climate-friendly transportation options, with showers, lockers, transit passes, financial rewards, and bike maintenance rooms.

- Provide resources for finding walking, biking, and transit routes or connecting with others to carpool, vanpool, or walk or bike together. Make carshare available during the day in case of unforeseen needs to pick up ill children or attend to other personal emergencies. (Many companies call this a "guaranteed ride home.")

- Convert your company fleet to electric vehicles, especially where renewable energy is available.

- Combine trips, and carpool for efficiency.

- Provide the time and flexibility employees need to bike or use transit to get to local meetings or appointments during the workday.

- Maintain vehicles to keep them running efficiently.

- Use the smallest vehicle that works for the job.

- Schedule video conferences and conference calls instead of travel when possible.

- Travel to conferences or regional meetings by train, car, or bus rather than airplanes whenever possible.

- Don't use a company jet. If you must fly, it's much more efficient to fly on a commercial airline.

Energy

Where you may find energy savings depends on the type of workplace, but nearly every worksite can be more energy-efficient.

- **Upgrade and maintain HVAC equipment.**

- Install renewable energy technology, such as solar, wind, or geothermal power, on site.

- Power down computers, printers, and other equipment when they're not in use. Use smart power strips to eliminate energy vampires.

- Use energy-efficient equipment in manufacturing, and keep it well-maintained.

- Use LED lighting, and install motion sensors where rooms are used intermittently.

- Conduct periodic energy audits, and publish the results.

Kitchens and events

Many workplaces have kitchens, cafeterias, or break rooms, and companies often host events and other occasions with food.

- **Minimize food waste.** Provide only what people will eat, and arrange for leftovers to be donated appropriately.

- Provide recycling containers with clear instructions about what is recyclable. Provide compost receptacles.

- Provide water coolers and washable glasses instead of plastic water bottles.

- Prioritize vegetarian and vegan options; minimize meat and cheese in meals and appetizers.

Financial investments

Just as you want your personal finances to support a healthy climate, companies and organizations should ensure their investments aren't doing harm.

- **Divest from fossil fuels in pension funds.**

- Offer socially responsible options for 401(K) plans.

- Divest from fossil fuels in all accounts; choose banks that don't finance the fossil fuel industry.

Purchasing

Use your company or organization's purchasing power to influence climate action upstream.

- **Source responsibly for supplies** such as paper, furniture, and office equipment.

- Source responsibly for raw goods for manufacturing.

Political action and policy

Businesses have political clout!

- **Lobby for climate-friendly policies and regulations** at the city, county, regional, state, and federal levels.

- Develop a climate action plan for your organization or company. Inform that plan with a climate audit.

- Urge your company to donate only to politicians and political groups that are actively working to curb climate change.

- Work with industry groups to improve industry-wide practices, policies, standards, and sourcing options. Ask employees for ideas and offer rewards for the ones the company adopts.

- Encourage employees to take action at home and in their communities. Support employee groups committed to climate action.

- Share strategies, successes, and lessons learned with other companies and organizations.

- Publicize your efforts and their effect to help others see what's possible and help change norms. True climate action is good press!

In local governments

In the United States, each of us lives in a community that is governed by policies, rules, and laws passed by members of that community and its elected leaders. You can make a difference by electing climate-conscious leaders (or running for office yourself!) and by letting your elected leaders know that you support climate-friendly policies.

As you advocate, you also have the opportunity and the responsibility to work to make those policies more equitable. Climate change shines a light on the inequities and injustices that have plagued us for too long. It's the communities that are most vulnerable who bear the brunt of the effects of climate change and, too often, the expense of policies designed to curb climate change. One of the most effective things you can do is to help to center the voices of those who have been disenfranchised, whose communities have not received as much investment, and who have been the victims of environmental pollution and degradation for decades or more.

Climate action plan

Every government—city, county, and state—should have a robust climate action plan that is updated regularly with community input, particularly from the most vulnerable populations, including poor communities, people of color, indigenous populations, and people with disabilities.

If your local government doesn't have a climate action plan, or it isn't strong enough, meet with your city or county council member or your state legislator to discuss the reasons you need a robust plan. Share model legislation with them. Then encourage others to reach out to their representatives too.

Sample climate action plans and guides to developing a plan are available at these sites (and many others):

www.c2es.org/content/state-climate-policy/

zeroenergyproject.org/advocate/model-climate-action-plans-ordinances/

www.epa.gov/statelocalenergy/local-action-framework

Transportation

Curbing emissions from transportation means minimizing the use of private vehicles and increasing walking, biking, transit, and carpooling. Electrifying vehicles can help, but they are not the ultimate answer—we need to reduce the number of vehicle miles driven, period.

There are several strategies to reduce the use of private vehicles and increase walking, biking, and transit:

- **Increase transit funding, reliability, and access.**

- Make transit free, especially for people with limited income.

- Charge congestion fees. Be careful to structure the program in a way that is equitable.

- Lower speed limits and build bike infrastructure to make biking safe and convenient.

- Support equitable bikeshare programs to make biking available to those who don't own their own bikes, and to provide spontaneous last-mile travel options.

- Repeal helmet laws, which not only deter riding but are often enforced inequitably.

- Maintain and improve pedestrian infrastructure. Pedestrian safety means full accessibility for people with disabilities, whether they have limited vision, use wheelchairs or walkers, have limited hearing, or have cognitive or other invisible disabilities.

- Repeal jaywalking laws, which are often enforced inequitably, and which prioritize vehicles over people.

- Charge for parking. Everywhere.

- Require employers to meet commute trip reduction goals; provide tax incentives for employers who exceed those goals.

- Require major institutions, such as hospitals and universities, to have robust transportation plans and to meet commute trip reduction goals.

- Make sure public buildings are conveniently accessible from transit, by bike, and by walking or rolling.

- Resist efforts to build or expand highways. Decades of experience have proven that expanding highways creates induced demand, leading to more people driving and not reducing congestion.

- For government or organizational fleets, use electric vehicles, and use them only when necessary. Expect government staff to use transit, bike, walk, roll, or use videoconferencing to attend meetings and events when possible. If you need to travel to other areas of the state or country, carpool or take a train or bus.

- Invest in and maintain an efficient fleet of emergency and maintenance vehicles.

Building codes/zoning

Denser neighborhoods are more likely to be walkable neighborhoods, especially when zoning mixes residential, commercial, and institutional uses. With walkable neighborhoods, you not only save energy, reduce vehicle miles traveled, and combine resources, but you build stronger and safer communities. There are several successful strategies for making neighborhoods more climate-friendly.

- **Make sure zoning allows for "missing middle" housing options** (between single-family homes and large apartment complexes) to increase density and provide more options.

- Zone for more density around transit hubs.

- Implement growth management plans with clear boundaries to avoid sprawl.

- Require buildings to meet LEED standards or better for energy efficiency.

- Ban natural gas in new construction. Provide technical and financial assistance for owners of existing buildings to transition from fossil fuels to electricity.

- Minimize or eliminate parking requirements.

- Consider impact fees.

- Provide financial support for home weatherization and other energy-efficiency measures, especially in low-income communities.

- Incentivize the installation of solar, wind, and geothermal energy systems for residential and commercial buildings.

Energy

City and state policies have a tremendous influence on the adoption of renewable energy and energy efficiency measures. Governments of all sizes can also reduce their own energy use.

- **Require that utilities provide a minimum percentage of renewable energy.** Increase that goal every five years.

- Encourage utilities to develop technology that lets them use electric vehicles as battery storage, so clean energy is available even when the sun isn't shining and the wind isn't blowing.

- Provide incentives for property owners to install solar panels; pay for solar energy production.

- Set minimum code standards that match LEED or better.

- Conserve energy in public buildings. As in homes, schools, and other workplaces, turn off equipment when it's not in use, and use the most energy-efficient equipment available.

- Lead by example: install solar panels, wind turbines, and geothermal energy systems in public buildings.

- Ensure that energy-related financial support is most available to people with the least income. Keep in mind that tax incentives typically benefit people with higher incomes.

Investments

Governments handle a lot of money. Make sure that money is working toward sustaining a livable climate.

- **Divest pension funds from fossil fuels.**

- Bank at socially responsible banks.

- Provide low- or no-interest loans for climate-friendly programs, especially for low-income residents.

Community-led resources

Governments can help community members support each other.

- **Provide matching grants** to help community members develop the projects that meet their needs.

- Support and promote community gardens.

- Support tool libraries, available to all neighborhoods.

- Work with community partners to coordinate work parties to rehabilitate parks, clean up waterways, and weatherize homes.

In healthcare

Climate change is a public health crisis, yet the healthcare sector contributes 10% of greenhouse gas emissions in the United States, with the majority from hospitals. Many healthcare providers now recognize that they have a responsibility to reduce the environmental effects of healthcare services while they care for their patients.

If you or a family member is a patient, ask your healthcare provider what they're doing to reduce climate emissions. Ask to see your provider's climate action plan.

If you are a medical professional or staff member at a healthcare institution, work from within to ensure you're moving toward carbon neutrality.

Energy use

Wasting energy increases carbon emissions and costs money. Hospitals and clinics can save energy in many ways.

- **Install LED lighting**, which can be adjusted to simulate daylight, improving patient and staff health and morale as well.

- Turn off lights that aren't needed; install motion sensors and timers where appropriate.

- Keep medical-grade computers on and ready for use, but install apps that let them sleep, using less energy until needed.

- Use non-electrical medical devices where possible.

- Install solar panels, wind turbines, and geothermal heat pumps to generate renewable energy.

- Monitor and adjust room air exchange frequency, based on whether operating rooms and patient rooms are in use.

Anesthesia

Anesthesia gases are greenhouse gases. Only about 5% of inhaled anesthesia is metabolized; the rest is vented into the atmosphere. Two common anesthesia gases, desflurane and sevoflurane, have very different climate effects. Desflurane is potentially twenty times as damaging as sevoflurane. Sevoflurane is also less expensive; Providence Health hospitals in Oregon made the switch to curb climate emissions and saved $500,000 in the process.

• If you're planning surgery, inquire about the anesthesia that will be used.

• If you're a healthcare professional, ask your hospital to use sevoflurane as its default for anesthesiology.

Transportation

Cutting transportation emissions is an obvious place to start for any institution or workplace.

• **Provide incentives for staff to commute to work on foot, by bike, using transit, or carpooling.**

• Design medical campuses to be welcoming and accessible to those who walk, bike, or use transit to travel there.

• Work with city and transit authorities to ensure there are safe, convenient walking and biking routes to your facility as well as frequent and reliable transit service.

• Where appropriate, encourage medical staff to meet with patients via telemedicine, eliminating transportation emissions and offering convenience to patients.

Plastics

About 25% of waste generated by a hospital is plastic. Hospitals routinely use single-use plastics because they're cheap, durable, and sterile.

But they contribute to microplastic pollution in oceans; require tremendous amounts of fossil fuels to manufacture, package, and ship; aren't typically recycled; and are potentially toxic to patients and staff.

- **Substitute safe, sterile, reusable equipment for at least some of that disposable plastic.** For example, choose rigid sterilization containers instead of polypropylene "blue wrap."

- Recycle—better yet, reuse—anything that hasn't had contact with patients, such as packaging and storage containers.

- As a patient, when you end a hospital stay, ask what will be thrown away. Scoop up the basins, plastic water pitchers, and anything else you can find a use for that would otherwise be discarded.

Cafeteria

Cafeterias provide opportunities for energy savings and sustainable practices.

- **Use energy-efficient appliances.**

- Prepare nutritious, local, sustainably grown food.

- Provide and promote meat- and dairy-free meals.

Buildings

Emissions can be reduced through green building design principles and the choice of materials.

- **Integrate daylight and natural ventilation** to reduce energy use and improve the health and wellbeing of patients and staff.

- Install a green roof or paint roofs a light color.

- Design medical facilities that exceed LEED standards.

- Install flooring that is sustainably manufactured, durable, and easy to maintain. Quiet floors help patients sleep well, improving recovery and shortening hospital stays.

Influence and investments

Medical institutions are often leaders in their communities. They can have broader reach through strategic investments, education, and advocacy.

- **Speak with elected officials and the press about the public health implications of global warming.**

- Divest from fossil fuel and other polluting sectors.

- Work with local elected and community leaders to accelerate climate action.

Resources

If you work for a healthcare institution, check out the following resources for more ideas, support, and guidance.

- Global Green and Health Hospitals (GGHH), a project of Health Care Without Harm (noharm.org).

- Health Care Climate Challenge (greenhospitals.net). More than 200 institutions representing more than 18,000 hospitals and healthcare centers in more than 30 countries have joined the challenge and committed to taking meaningful action.

- *Climate Action: A Playbook for Hospitals* from Health Care Climate Council at climatecouncil.noharm.org includes great examples of hospitals taking action.

- Practice Greenhealth (practicegreenhealth.org) is a nonprofit that works to make hospitals more sustainable.

- "Climate-smart health care: Low-carbon and resilience strategies for the health sector" is a report from the World Bank available at documents.worldbank.org (It's intended for providers in developing countries, but the information is useful worldwide.)

A climate-friendly afterlife

You're making conscious choices to reduce your effect on the climate while you're alive. But what happens when you die?

In nature, bodies often rot where they lie if they're not consumed by other bodies. But humans have other ideas. After-death rites and practices vary around the world; some are more climate-friendly than others, and some more supportive of community as well. In the U.S., we've mostly had two options: burial in the ground or in a crypt, or cremation. The first takes up precious space and often leaches embalming chemicals into the ground. Cremation reduces the body to ashes, but spews fossil fuels into the air in the process.

Personally, as a long-time organic gardener who loves to play with my compost pile, I'm excited about a new option in my home state, Washington. Terramation composts the body through an optimized process that takes about two months. The result is about a pickup truck's worth of garden-ready compost from the body and the grasses and other organic materials mixed with it.

Colorado has also legalized terramation, and other states may follow. If you'd like your body to be composted when you die, let your state legislators know. Once it's legal in your area, include instructions in your will and let your loved ones know. If you have suggestions for what to do with the compost, share those, too. (Most facilities will probably be able to send compost to parks if your loved ones don't need a truckload of compost.)

While we're thinking about death, consider burying pets instead of cremating them. Check your local regulations, but in most places, you can bury an animal on property you own as long as the grave is deep enough and far enough from a waterway.

Parting thoughts

The world as we have created it is a process of our thinking. It cannot be changed without changing our thinking. - Albert Einstein

The best remedy for hopelessness is action. If you feel overwhelmed by the climate crisis, or alone in your fear and anxiety, remember that everyone on this planet is facing an uncertain future. You have the power to make responsible choices that will make a real difference. You can stand up with others to demand the changes from industry and government that will keep our world livable, and you can hold leaders accountable to create a just transition to a clean-energy society.

Working together, we can change our individual lives and our institutions to preserve the livability of this planet's climate while increasing equity and improving the health of our communities.

As we address the climate crisis and any other emergencies that arise, we have the opportunity and the responsibility to center the communities that are most affected. Too often, communities of color, low-income communities, and other under-resourced groups bear the brunt of the harm, while more privileged communities continue to exploit the world's resources and churn out greenhouse gases without thinking about the harm they (we) do.

Commit to rethinking your habits and routines and encourage others to do the same. While I've offered many examples of actions you can take, the ultimate goal is to be more conscious of our impact upon the climate, the environment, and each other. If we move through the world more mindful of what we use, what we waste, and who is harmed or benefited by our actions, we can build a better world together.

Notes

I will do these things now: _____

_____

_____

_____

_____

_____

_____

_____

_____

_____

_____

_____

_____

Notes

I want to learn more about: _____

_____

_____

_____

_____

_____

_____

_____

_____

_____

_____

_____

Notes

I'll share these ideas: _____

_____

_____

_____

_____

_____

_____

_____

_____

_____

_____

_____

_____

_____

Gratitude

I am deeply grateful to Cynthia Scheiderer for reviewing multiple drafts and helping me find better, clearer ways to express ideas. Tonia Larsen provided a critical eye and consistent encouragement. Barb Radin helped me think about my audience and cheered me on. I owe a huge debt to Nivi Achanta, founder of the Soapbox Project, for her insights and expertise.

I can't even begin to name all the people I've talked with about the book and the ideas in it. Countless people shared the aspects of climate action they're most passionate about, what confuses them, and how they think a book like this might be used. Each conversation that sent me home to scribble madly made this book a better one.

None of the ideas or information in this book started with me. They are culled from newspaper and magazine articles, books, scientific papers, and websites documenting the thorough, important work done by amazing scientists, journalists, and advocates. We all owe them a huge debt, and any errors in this book are my own.

I am deeply grateful to the indigenous Water Protectors and other committed activists who are putting their bodies on the line to preserve a livable earth for humans and other species. Their actions are inspiring millions to step up and do our part.

Finally, I want to give a shout out to my mother, from whom I inherited incorrigible optimism. Thanks, Mom!

CPSIA information can be obtained
at www.ICGtesting.com
Printed in the USA
BVHW032143040822
643866BV00006B/102

9 781734 864809